Let's Look at Lizards

GILA MONSTERS

By Seth Lynch

Gareth Stevens
PUBLISHING

Please visit our website, www.garethstevens.com. For a free color catalog of all our high-quality books, call toll free 1-800-542-2595 or fax 1-877-542-2596.

Library of Congress Cataloging-in-Publication Data
Names: Lynch, Seth, author.
Title: Gila monsters / Seth Lynch.
Description: Buffalo, New York : Gareth Stevens Publishing, [2025] | Series: Let's look at lizards! | Includes index.
Identifiers: LCCN 2023034774 | ISBN 9781538292730 (library binding) | ISBN 9781538292723 (paperback) | ISBN 9781538292747 (ebook)
Subjects: LCSH: Gila monster–Juvenile literature.
Classification: LCC QL666.L247 L96 2025 | DDC 597.95/952–dc23/eng/20230821
LC record available at https://lccn.loc.gov/2023034774

First Edition

Published in 2025 by
Gareth Stevens Publishing
2544 Clinton Street
Buffalo, NY 14224

Editor: Brianna Propis
Designer: Leslie Taylor

Photo credits: Cover Vaclav Sebek/Shutterstock.com; p. 5 Evelyn D. Harrison/Shutterstock.com; p. 7 reptiles4all/Shutterstock.com; p. 9 (top) Evelyn D. Harrison/Shutterstock.com, (bottom) Vaclav Sebek/Shutterstock.com; p. 11 Yoshan Hasintha/Shutterstock.com; p. 13 Kris Wiktor/Shutterstock.com; p. 15 Vaclav Sebek/Shutterstock.com; p. 17 Vladislav T. Jirousek/Shutterstock.com; p. 19 David Clode/Unsplash.com; p. 21 Sibons photography/Shutterstock.com; p. 23 MonsterDoc/https://en.m.wikipedia.org/wiki/File:Gila-monster-6-eggs.jpg.

Printed in the United States of America

CPSIA compliance information: Batch #CSGS25: For further information contact Gareth Stevens, New York, New York at 1-800-542-2595.

Contents

Gila monsters are lizards.
They are large.

They have scales.
These feel bumpy.

They are dark gray.
They have spots.
Spots can be orange,
pink, red, or yellow.

They have a big tail.
It helps them store fat.

They live in the desert.
They live in the ground.

They come out to eat.
They come out to lay
in the sun.

They eat mice and birds.
They eat eggs.

They cannot see well.
They find food by smell.

Their teeth are strong.

Mothers lay eggs.
They keep the eggs safe.

Words to Know

desert

egg

scales

Index